Construction Technology

The How, What, and Why

Epris E. Ezekiel

Copyright 2024© Epris E. Ezekiel

All rights reserved. This book is copyrighted and no part of it may be reproduced, distributed, or transmitted in any form or by any means, including photocopying, recording, or other electronic or mechanical methods, without the prior written permission of the publisher, except in the case of brief quotations embodied in critical reviews and certain other non-commercial uses permitted by copyright law.

Printed in the United States of America
Copyright 2024© Epris E. Ezekiel

Contents

Introduction 1

Chapter 1 2

What is Construction Technology? 2

Chapter 2 8

Addressing labor issues in construction 8

Chapter 3 17

Modern construction technologies are revolutionizing the sector. 17

Chapter 4 25

Building information modeling (BIM): Uncovering its potential 25

Chapter 5 28

Choosing Construction Technology That's Right for You 28

Chapter 6 35

Innovative materials: Revolutionizing the building sector 35

Chapter 7 43

Using augmented reality to visualize massive physical projects 43

Conclusion 53

Introduction

The construction industry has faced many obstacles in recent years. These include labor shortages, worksite injuries, inflationary competition, and project delays that hurt construction companies, workers, and clients. It is already difficult to complete safe, high-quality work, find skilled workers, adhere to strict budgets, maintain project collaborations, and meet deadlines. Numerous building firms are looking for sustainable, scalable solutions in response to these issues. Numerous obstacles impeding the construction business are addressed by construction technology, in all of its manifestations.

We will explore construction technology in its entirety in this article, including its definition, the advantages it offers construction workers, real-world examples of its application, and tips for putting its many solutions into effect.

Chapter 1

What is Construction Technology?

These programs and instruments are designed to simplify the trickiest and riskiest parts of construction work, not to take the place of knowledgeable laborers. Planning, designing, building, and managing construction projects are all accomplished with the help of various tools, hardware, and software that are part of today's construction professionals' metaphorical toolbox.

The pre-construction process has been made simpler by construction technology, in addition to the tools, machinery, and software used to support building activities. Construction managers and their teams may get started on every phase of a project on the right foot by putting in place tools for bid management, document organization, and teamwork.

The construction industry has benefited greatly from the introduction of mobile project management apps, smart tools and equipment, digital schematics, autonomous machinery, and 3D printing technology that produces high-quality, reasonably priced building materials in advance.

What are the Benefits of Construction Technology?

The development of construction technology has received about $10 billion in funding over the last ten years from both technology businesses and construction corporations. The way construction workers perform has changed dramatically in recent years because of technological improvements in the field. Let us examine the principal advantages that construction technology offers to the construction sector as a whole.

- ❖ **Cost Reductions:** Building firms and their clients can save money by using building technologies to cut waste and maximize resources. Because building technology has allowed for a 30% reduction in expenses from 6% to 30%, 26% of construction executives have increased the usage of prefabricated and modular materials in their organization.
- ❖ **Enhanced Cooperation:** Team members and supervisors can see real-time updates, communications, and more by using mobile project management software that is cloud-based. While on the job site or off, construction professionals may stay informed about the most recent project announcements thanks to an intuitive project management dashboard. Additionally, cross-trade conflicts are minimized and cooperation is made

possible by a completely automated robotic layout, which prints each trade's layout from the coordinated model.

❖ **Adherence to Compliance:** Information may be easily recorded, signed, and sent to regulatory agencies with the help of tools like document management, photo and video storage, and real-time notifications. By following the laws and regulations that control the work being done, construction teams may make sure that their businesses and projects continue to operate legally.

❖ **Increased Quality:** To help workers do challenging tasks with more accuracy, advanced technologies like augmented reality (AR) overlay instructions directly onto a job site. High-quality and precise work is also ensured by robotic layout, which prints the synchronized digital model directly on the surface with a 1/16" accuracy. With the use of a tool like this, building quality is increased overall and error risk is decreased.

- ❖ **Enhanced Effectiveness:** Technology in construction streamlines procedures and shortens construction timelines. While certain uses, like 3D printing, produce building components more quickly, others, like robotic layout printers, may print digital building models straight onto the surface of a construction site. With the help of these thorough schematics, installation, and layout can be completed more precisely, reducing the need for rework and conserving time and materials.
- ❖ **Enhanced Safety:** Construction sites can be made safer by utilizing construction technology. To improve safety procedures

on the job site, construction managers can, for starters, examine construction safety data. To safeguard human workers, construction companies have also begun to use drones to investigate potentially hazardous regions. Some businesses have begun utilizing wearable technology—such as temperature sensors, motion detectors, and carbon monoxide detectors—to monitor worker safety. Furthermore, robotic layout systems automate the construction layout's most repetitive and injury-prone tasks, proactively eliminating risk from the job site.

Chapter 2

Addressing labor issues in construction

There is still a severe and pervasive labor shortage in the construction sector. According to a survey conducted by the Associated General Contractors of America (AGC), 91% of construction companies say they have trouble filling open positions, which raises prices and delays projects. To lower that number, it will be important to prepare, recruit, and reskill the workforce of the future. Technology can help close the skills gap.

Regarding the workforce deficit, how might technology be helpful? Initially, by making investments in innovative methods, more young persons from tech-savvy generations can be drawn to and remain in construction jobs. Secondly, labor requirements can be simplified by technology. A contractor may more effectively manage productivity levels and its personnel by implementing tech strategies to assist cut down on on-site worktime. Examples of these strategies include automating inventory and ordering processes and utilizing virtual construction technologies like Building Information Modeling (BIM).

Taking care of risk and safety in construction
To increase safety and internally manage workflows like estimating project scheduling, and project management, construction companies are turning to technology solutions, according to the JBKnowledge 2021 Construction Technology Report, which states that 44% of today's contractors have dedicated IT departments. Here are a few examples of how technology is enhancing productivity and security:

Telematics

With telematics, data on safe driving practices and driving habits can be collected via phone apps, cameras, AI sensors, and seatbelt monitors. Smart business decisions are made by successful organizations in every industry, including construction, by utilizing technology and data. Employers may make data-driven decisions and safeguard their workforce, fleet, and financial line by investing in telematics.

Accessories

With wearable technology becoming more and more popular in consumer markets, the construction sector is now thinking about how to integrate various capabilities to increase safety on the job site. Wearable technology has several possible uses, including tracking a worker's movements or health and giving immediate alerts if unsafe working conditions arise. For instance, smartwatches can identify alertness and weariness in drivers and equipment operators, while other wearables can track physiological reactions and warn employees about heat stress to lower the risk of heat-related illnesses.

Mobile gadgets

The JBKnowledge analysis demonstrates the exponential increase in the significance of mobile capabilities, with 85% of the organizations polled stating that mobility is "important" or "very important" to their operations, up from just 59% who felt the same way ten years ago. Of those asked, 91% said they regularly use their smartphones for work-related tasks.

The building process is rapidly becoming more efficient with the use of smartphones, tablets, and other portable devices, from drawing and approving designs to monitoring the work after it has been completed.

The adoption of mobile technologies can provide several benefits for safety, including speedier incident reporting and injury documentation, real-time communication with all members of a construction team, and a decrease in the risk of injuries and workers' compensation claim costs.

Labor-saving techniques that increase worksite productivity

Not only can automation provide precision, speed, and accuracy in some construction jobs like material handling, packing, and welding, but it can also save a substantial amount of time and money. These three tactics will increase accuracy and save time for businesses, increasing the likelihood that projects will remain on schedule and under budget.

Robotics

With important advantages like these, robotics is continuing to alter several parts of the construction industry.

- ✓ Accelerated construction
- ✓ Fewer injuries and safer workplaces
- ✓ Reduced waste from construction sites

These advantages may also result in cheaper building, financing, and insurance expenses from a financial perspective.

In particular, robotics can help trained workers become more productive by allowing them to supervise projects more closely and collaborate with robots to complete tasks like bricklaying, demolition, and drywall installation. Similar to exoskeletons, corobotics can lessen the risk of musculoskeletal disorders associated with the workplace by minimizing worker fatigue from lengthy overhead work and lifting, lowering, and carrying building materials.

"Skilled workers can spend more time managing the project and collaborate with robots to become more productive," according to Robotics.

The construction robot market in the United States is predicted to increase at a compound annual growth rate of 12.9 percent between 2023 and 2028, even though adoption is still relatively limited. The process of creating objects with computer-aided designs (CADs) and 3D printers to produce materials that may be used to build buildings is known as 3D printing, and it is anticipated to have a substantial market share

Flaws in construction

One important strategy to reduce building flaws is to use technology. Contractors can integrate AI into their workflows as they grow more advanced, perhaps reducing the likelihood of expensive claims.

Engineering for climate resilience

Contractors and city planners have begun investing in resilient construction, which is the process of creating stronger structures and infrastructure to shield cities and individuals from extreme weather and other calamities, in response to a world where weather patterns are becoming more unpredictable. Businesses can strengthen their initial construction and adapt their current facilities by utilizing analytics to assist in anticipating the implications of climate risk.

Water intrusion apparatus

Delays, errors, mold, pollution, and other issues can arise from water incursion. You can save time and thousands of money by ending it sooner rather than later. Water intrusion technology is becoming increasingly widespread in the building sector, and cloud-based and Internet of Things technologies have removed infrastructure-related costs and implementation constraints.

Chapter 3

Modern construction technologies are revolutionizing the sector.

Construction technology is one industry that is being drastically changed by the digital revolution. It is now more than just a change; it is an essential component of contemporary building projects. Using cutting-edge building technologies that optimize workflows, cut waste, and boost productivity is part of this shift.

This industry is changing quickly. Our reliance on bricks and mortar is no longer sufficient; instead, a world of opportunities has been made possible by current construction technologies. These developments transform the way we construct and design buildings, from robotics to digital tools and software integrations.

Adopting this cutting-edge building technology might be difficult, though. Overcoming obstacles like resistance to change or apprehension about upsetting established workflows is typical.

What are the newest technologies used in building?

Building information modeling (BIM), cloud-based project management, robotics, drones, thin-jointed masonry, and self-healing concrete are some of the cutting-edge technologies that are revolutionizing the construction industry. Other innovations include 3D printing, augmented reality (AR), modular building, and mobile technologies.

Modern construction encompasses a variety of approaches that combine traditional methods with technological innovations. These include lean construction, which minimizes waste, green buildings that use sustainable practices, off-site prefabricated or modular structures, smart homes that integrate IoT devices, and micro-homes that maximize small spaces.

Which four categories of technology are there in construction?

The four primary categories of contemporary construction technology are enhanced building materials, automation & robots, building modeling & design software, and information technology for project management. Each kind is essential for raising production and efficiency.

Digital technologies that simplify procedures, cut waste, and enhance decision-making are a hallmark of modern construction technology. Real-time data access, enhanced cloud-based collaboration via big data integration, predictive analytics, and the deployment of automated equipment and robotics are all included.

Using Technologies for Cloud-Based Project Management

Cloud-based project management solutions have become indispensable for effectively organizing complicated activities in today's fast-paced construction industry, where teams use a wide variety of materials. These platforms offer real-time updates, which let teams collaborate better and make decisions more quickly, which increases productivity.

These technologies not only increase efficiency but also dramatically lower project costs by better-allocating resources.

Construction management software that runs on the cloud has several advantages.

1. **Eco-Friendliness:** Additionally, cloud-based construction management software is a more environmentally responsible

option because it digitizes data and eliminates the need for paper.

2. **Information Security:** The confidentiality of all project-related data is guaranteed by the strong security features these systems offer, such as data encryption and secure user authentication.

3. **Scalability:** Cloud-based software offers prospects for sustainable corporate growth since it can quickly scale to accommodate more users and higher data volume as project requirements rise.

4. **Increased Awareness:** Better project control can result from project managers' ease of tracking modifications, keeping an

eye on project progress, and spotting possible bottlenecks.

5. **Enhanced Risk Handling:** These platforms give teams access to instruments for risk assessment and mitigation, assisting them in seeing possible problems before they become serious ones.

6. **Economy of Cost:** By removing the requirement for physical storage space and the associated costs of manual data entry and record-keeping, they provide an affordable alternative.

7. **A rise in productivity:** Several repetitive operations can be automated by cloud-based construction management

software, which saves a significant amount of time and boosts productivity.

8. **Enhanced Cooperation:** By offering a central home for all project-related papers, drawings, and schedules, these systems help teams collaborate more effectively.

9. **Instantaneous Information Access:** Real-time access to pertinent project information is made possible by the cloud-stored data. Mission-critical data is immediately available as a result, facilitating prompt decision-making.

10. **Streamlined Interaction**: No matter where they are in the world, cloud-based software enables instantaneous, seamless communication between all members of a construction team, including architects

and site workers. This makes sure that everyone is aware of the most recent advancements in the project.

Along with revolutionizing the construction sector through increased productivity, lower costs, and improved communication, the use of cloud-based project management systems is also supporting environmental sustainability.

Chapter 4

Building information modeling (BIM): Uncovering its potential

With digital technologies like BIM playing a critical role, the construction sector is transforming. Better more, BIM simplifies decision-making throughout the project, not just when producing visually appealing 3D models. The digital representation of a facility's functional and physical elements is correct thanks to this new development in construction technology. Contractors, engineers, and architects may foresee possible problems before they become expensive ones with the help of this indispensable tool.

It also contributes to waste reduction and increased sustainability by monitoring a project's whole lifecycle from design to completion. BIM is utilized for everything from facility management and infrastructure planning to safety planning and fire prevention, so the ramifications are extensive.

Using predictive analytics with large data in the construction industry

Big data integration may dramatically increase productivity in all aspects of your business, from strategy to execution. With this powerful resource at their disposal, firms may decide to use up-to-date information.

- ❖ **Mobile technology improves efficiency & visibility.**

    It is now easier and more efficient than ever before to track resources and personnel in several places using real-time data made possible by mobile technologies. With the help of these technologies, we can always be instantly aware of everything going on at the job site, assuring maximum efficiency.

- ❖ **Utilizing previous patterns to forecast results**

  Professionals in our field may reliably predict future results thanks to predictive analytics, which leverages historical patterns extracted from large databases. For example, we might anticipate any delays caused by inclement weather by comparing historical weather data with work schedules. With large-scale projects where numerous factors necessitate regular monitoring, this level of foresight guarantees smoother overall operations in addition to saving significant resources.

Chapter 5

Choosing Construction Technology That's Right for You

Making the right construction technology choice for your organization can be challenging for decision-makers due to the abundance of options available.

Thankfully, choosing the finest building technology for your particular business needs should be simpler with the help of the following suggestions. To make the best-informed choice possible, take into account the following.

Your Goal

Examine the primary difficulties and obstacles that your construction company faces before making any purchases. Your primary goal for this new technology will be informed by your identification of these pain locations, which will also assist direct the tool selection process. Begin by posing the following queries:

- ✓ What is it that the construction crew is always complaining about?
- ✓ Is constant rework an issue?
- ✓ Do trade tensions arise during layout?
- ✓ What frequently leads to delays in projects?
- ✓ Where is the company's biggest inefficiency?

Your Essential Components

You now know the main issues that your team is facing and the purpose of your new solution, therefore the next step is to determine which features will help you reach your target. You must identify the features that are essential to your team, the features that you don't need, and the ones you would like to know more about. Not every feature will be significant to your team. Determine which features best fit your current tech stack and company requirements by evaluating the offerings that each construction tech solution promotes.

- ✓ Will employing augmented reality be advantageous for your business?
- ✓ Does your group use software for digital modeling?
- ✓ Which procedures stand to gain the most from automation?

Budget for the Company

The adoption of new technology is significantly influenced by cost. Examine the costs associated with the pain problems you are trying to tackle as well as the amount of money your organization can afford to invest.

It's also critical to comprehend the billing procedures of the software you are choosing.

- ✓ Do different options for packages?
- ✓ Is there a one-time fee associated with the technology?
- ✓ Do subscriptions charge on a monthly or annual basis?

Usability of the Product

If your team is unable to learn how to use every feature of the system, then perhaps an innovative piece of technology is not worth the money and effort invested. The best implementation outcomes will come from selecting a technological solution that is simple to use and can be rapidly grasped. In this manner, your staff will be able to utilize the product effectively without getting impatient, squandering time, or avoiding the software entirely.

- ✓ Can employees with varying degrees of experience efficiently utilize the capabilities of the software?

- ✓ Does the software integrate with existing tools and programs?
- ✓ Does this technology include instructions, a handbook, or a demo?

Construction's shift toward robotics

The goal of integrating robotics into construction is not to replace human labor, despite common opinion to the contrary. Enhancing efficiency and safety on the job site is the main focus instead.

Replicated tasks that eventually become physically burdensome for humans are being taken on by robots. Furthermore, they are demonstrating their value by functioning safely in dangerous situations where the presence of humans would present serious concerns.

Drones: A fresh viewpoint on building projects

Drones have emerged as a key element of contemporary construction technology in recent years. With their unparalleled aerial viewpoint and constant progress tracking, these unmanned aerial vehicles provide priceless data through thorough site surveys.

These aerial robots, equipped with sophisticated cameras and sensors that can take clear photos or movies from previously unreachable positions, are completely changing the way we view our workplaces. Industry insiders estimate that the use of drones may save billions of dollars in costs in several industries, including construction.

Industry sources claim that mobile technologies go beyond simple graphics and offer real-time data visibility, making it possible to follow resources—including drones—efficiently. This feature helps project managers track progress remotely against initial plans and mitigate risk factors related to delays and budget overruns by identifying possible problems early on. In the end, this greatly increases the efficiency of project management overall.

Chapter 6

Innovative materials: Revolutionizing the building sector

The building industry is not behind when it comes to cutting-edge materials. We are seeing a revolution in this industry as conventional materials give way to more sophisticated substitutes.

Self-healing concrete has proven to be the most remarkable of these new developments. Using bacteria, this clever substance automatically fixes cracks that develop over time, a common problem with regular concrete. Apart from self-healing concrete, the construction sector is being shaped by many novel materials:

1. **Biomimetic materials:** These materials, which draw inspiration from nature, imitate the chemical or physical properties of natural materials. Seashell

concrete, a substance that imitates the sturdiness and resilience of seashells, is a prime example.

2. **Graphene:** Graphene is a "wonder material" because it is extremely strong, flexible, and light at the same time. Its application in building is still in its infancy, but it has the potential to help with water purification and build stronger, lighter structures.

3. **Timber with cross-laminations (CLT):** CLT is a much lighter substitute for conventional building materials and has comparable strength properties to concrete. Because of this, it's a great option for construction projects that care about the environment.

4. **Aerogel insulation:** Aerogel, sometimes known as "frozen smoke," is a very light substance with remarkable insulating properties. Using aerogel as insulation in buildings can save a lot of money on energy throughout the structure.

5. **Fiber carbon:** Carbon fiber is being used more and more in the building industry because of its low weight and great tensile strength. It is frequently utilized as a reinforcing material, particularly when building bridges and modifying ancient buildings.

6. **Transparent Aluminum:** Known by another name, aluminum oxynitride, this new substance is nearly as clear as glass and incredibly strong. It can be used in

construction for transparent structural elements or even for bulletproof windows.

Reusing recyclables to promote sustainability

- ✓ Recyclable materials are increasingly being used in building projects as a way to support worldwide efforts to meet sustainability targets.
- ✓ Reinforcement bars made of recycled plastic
- ✓ Architectural concepts incorporating reclaimed wood
- ✓ Glass bottles transformed into ornamental pieces.

These revolutionary concepts are radically changing and remaking our built environment; they are not just eco-friendly solutions. They are laying the foundation for a more sustainable and greener future with their creative methods and eco-friendly behaviors.

Construction using 3D printing is the way of the future.

It is impossible to ignore the transformative impact of 3D printing when discussing the future of construction technology. This engineering marvel has expanded the industry's possibilities for mass production.

In this method, materials are deposited layer by layer to create three-dimensional objects from digital models. With the unparalleled degree of creative freedom, it offers, intricate geometries that were previously thought to be unachievable using conventional techniques can now be realized.

Navigating probable next obstacles

There will undoubtedly be challenges as we continue to explore the prospects presented by these cutting-edge technologies, but given their benefits, it is easy to see why so many businesses are eagerly implementing them into their operations now. Here are some possible difficulties that additive manufacturing may face in the future.

Making intricate architectural designs
Fabricating elaborate architectural designs is one area where this novel technology particularly shines. Such intricate constructions are often very time- and resource-consuming to produce using traditional methods; however, with the accuracy and efficiency that 3D printing offers, these are now more feasible. See how this idea is being used by architects here.

In addition to expediting production, it also drastically lowers waste generated during fabrication, which is a crucial benefit given the environmental effects of global construction projects.

Cutting-edge large-scale structures

Subsequently, full-scale buildings have been successfully printed using specially blended concrete compositions, going beyond individual parts or sculptures. Homes such as this one serve as examples of what the widespread adoption of new building technology could soon bring about in global housing markets.

These achievements demonstrate the revolutionary potential that 3D printing holds for construction projects in the future—they represent the realization of enormous concepts.

Cutting-edge large-scale structures

Subsequently, full-scale buildings have been

successfully printed using specially blended concrete compositions, going beyond individual parts or sculptures. Homes such as this one serve as examples of what the widespread adoption of new building technology could soon bring about in global housing markets.

These achievements demonstrate the revolutionary potential that 3D printing holds for construction projects in the future—they represent the realization of enormous concepts.

Chapter 7

Using augmented reality to visualize massive physical projects

The creative incorporation of augmented reality (AR) into construction operations is where the industry's future is found. With the use of this technology, large-scale physical projects may be managed and visualized in a novel way before any work is done, increasing productivity and decreasing mistakes.

In essence, these technologies are transforming how we plan and execute our projects by enabling new levels of insight and control over numerous elements.

Utilizing on-site augmented reality

Employees using mobile applications with augmented reality (AR) capabilities through tablets or smart glasses have advantages even after project planning is complete (source). With the help of these tools, you can minimize errors and improve overall quality by seeing precise component placement through useful data overlays.

In the future, construction processes may explore fascinating new vistas thanks to technological innovations like mixed reality, which combines aspects of virtual reality and augmented reality.

Using augmented reality to plan projects in a new way
One area where AR has advanced significantly is project planning. Architects and engineers have discovered a useful tool that helps stakeholders better comprehend design concepts early in the development process: accurate representations

made with interactive 3D models. Click this link to learn more.

1. Not to mention, it's helping construction teams use a variety of materials and operate in various places to collaborate better.
2. As a result of the early incorporation of feedback, costly redesigns are reduced.
3. It also guarantees that everyone engaged is aware of the goals for the build phase.

Modular building: An innovative approach

Modular building is becoming a disruptive factor in the construction sector. Building components are created in controlled settings off-site and subsequently assembled on-site in this method.

This novel approach to mass production guarantees uniform quality control for every component of the structure while simultaneously shortening project schedules.

The benefit of efficiency

The efficiency of this contemporary construction technology is one of its main selling points. Project completion times can be accelerated by up to 50% when components are built concurrently on- and off-site, as opposed to using traditional methods. Prefabrication eliminates the need for weather delays, which are a common cause of construction delays in conventional projects.

In addition to saving time, this system has the following important advantages:

1. **Security:** A smaller workforce on the job site lowers possible risks, resulting in generally safer working conditions.

2. **Durability:** Compared to typical processes, less waste material is produced during manufacture thanks to the application of precision engineering techniques.

3. **Control of quality:** Strict controls are used at every stage of the module manufacturing process to guarantee that excellent standards are upheld throughout the build cycle.

Taking on possible obstacles head-on

Adopting modular construction methods presents its own set of problems, much like implementing any new technology. For example, the logistics of transportation could be challenging based on the weight or size of prefabricated modules.

Before making a complete commitment, businesses adopting such approaches must carry out in-depth feasibility assessments, taking into account things like the local transportation infrastructure. This will help them get past these obstacles.

We will always need to be flexible to move toward more sustainable and effective methods but with the variety of materials accessible today combined with mobile technologies, The efficiency and affordability of modern construction technologies, especially modular construction, are reshaping the sector. Building parts off-site and assembling them on-site allows projects to be finished more quickly without sacrificing safety or quality. But it's crucial to carefully handle any difficulties like the logistics of transportation.

Using mobile technology to its full potential for improved visibility

The building sector is changing as a result of mobile technologies. With their ability to provide a clear perspective of real-time data and enable effective tracking across several locations, they are quickly emerging as important tools.

There are numerous benefits for construction project managers associated with this new wave of digital transformation. These apps are now necessary for increasing productivity, from tracking progress while on the go to accessing important papers from anywhere. Here are some ways that mobile technology is changing the construction sector.

Using cloud-based systems to facilitate real-time communication

In addition to location-based services provided via GPS integration, cloud-based platforms provide real-time team member engagement beyond geographical borders. Decision-making procedures move forward much more quickly because of this smooth communication network, which eventually results in increased productivity and lower project costs. Click here to find out more about using cloud computing in the context of contemporary building projects.

1. **Safeguard critical information:** The likelihood of losing crucial files as a result of human error or device malfunction is almost zero because every piece of data is automatically backed up onto safe servers. This makes them the perfect

option for handling complicated activities like those seen in today's dynamic environment.

2. **Encourage efficient dialogue:** Enabling instantaneous updates sharing with all stakeholders, including architects, engineers, and subcontractors, not only saves time but also averts misunderstandings, resulting in more efficient operations overall.

3. **Allow for immediate access:** Since everything is safely saved online and is always accessible via smartphones or tablets, everyone engaged is certain to be informed of any updates, even when they are not in the office.

Cloud-based systems, GPS tracking, and mobile

technology are revolutionizing modern buildings. These solutions are essential for increasing productivity and cutting costs because they preserve critical information, improve asset management, enable instantaneous cooperation across regions, and provide real-time data insight.

Conclusion

With its many advantages and ability to drive important trends that will influence project delivery in the future, construction technology, or Contech, is completely transforming the construction sector. Adopting construction technology has become more and more crucial for companies looking to stay competitive in today's market, from raising productivity and efficiency to strengthening safety and sustainability.

Predictive analytics, big data, and building information modeling (BIM) are not just catchphrases but revolutionizing productivity. Robots assist humans with dangerous situations and tedious jobs rather than taking the place of human labor. For monitoring and surveying purposes, drones have also found a place on-site.

Sustainable development is aided by cutting-edge materials like thin joint masonry and self-healing concrete. We have seen firsthand how augmented reality helps visualize projects before work begins, modular construction may expedite completion timeframes without sacrificing quality, and mobile technology can enhance site visibility. Some examples of these innovations include 3D printing and mobile technologies.